Trending Research Technologies in INFORMATION TECHNOLOGY

Authors

Dr. Neha Agrawal

Meenu Garg

Integrated Publications ™

New Delhi

Published By: *Integrated Publications* ™

Integrated Publications
H. No. - 3 Pocket - H34, Sector - 3,
Rohini, Delhi-110085, India
Email – info@integratedpublications.in

Authors: *Dr. Neha Agrawal and Meenu Garg*

Edition: *1st*

Publication Year: *2023*

Pages: *50*

ISBN: *978-93-5834-008-2*

Book DOI: *https://doi.org/10.22271/int.book.344*

Price: *₹523/-*

Contents

Chapter - 1

Emotion Detection Using Convolution Neural Network and Open CV in Real-Time

Authors

Neel Kamal
Department of IT, Maharaja Agrasen Institute of Technology, Delhi, India

Aakash Singh
Department of IT, Maharaja Agrasen Institute of Technology, Delhi, India

Neha Agrawal
Department of IT, Maharaja Agrasen Institute of Technology, Delhi, India

Meenu Garg
Department of IT, Maharaja Agrasen Institute of Technology, Delhi, India

Chapter - 1

Emotion Detection Using Convolution Neural Network and Open CV in Real-Time

Neel Kamal, Aakash Singh, Neha Agrawal and Meenu Garg

Abstract

Applications built by integrating Flutter and machine learning are applications that predict or manage stress. Its sole and primary function is to recognize the stresses and pressures facing people today and recommend ways to deal with them. Humans are constantly activated by various environmental stimuli that go unnoticed. In some cases, these stressors cannot be detected. But machines can handle this well, so you can write programs to make decisions on the fly. Today, many scientists and researchers rely on information gathered on the Internet to make the right decisions. This application also uses image recognition technology and camera data to determine if a person is stressed. This software takes an image with a camera, processes it, decides what to do with it and displays it on the screen. This application has a separate section that suggests different ways to manage stress.

Index terms: Stress prediction, facial recognition, machine learning, stress relief.

1. Introduction

Stress is a common or natural feeling that often occurs when a person is unable to cope with or complete certain situations or tasks. However, stress can become an incurable disease for anyone if ignored. Relationships, workload, financial difficulties, and other issues can contribute to this situation. It is often observed that anything that poses a threat or challenge to a person's well-being can lead to stress. Everyone can be stressed, and it's even essential to survival. Each has a set of reflex actions that help determine how to respond to potential threats. However, when the body is vulnerable, or when multiple stressors are present at the same time, it can be dangerous and negatively impact physical and mental health. Recognizing stress early is important to prevent it from spiralling out of control. Stress can negatively affect your daily life and cause various interpersonal and relationship problems. As a result, we developed a project that uses machine learning to detect stress from facial expression sand provide different coping mechanisms. The proposed framework identifies stress-related facial expressions such as anger, fear, and sadness, among others. Emotional states fall into two categories as shown in Figures 1 and 2.

Fig 1: Different emotion states which are stressed

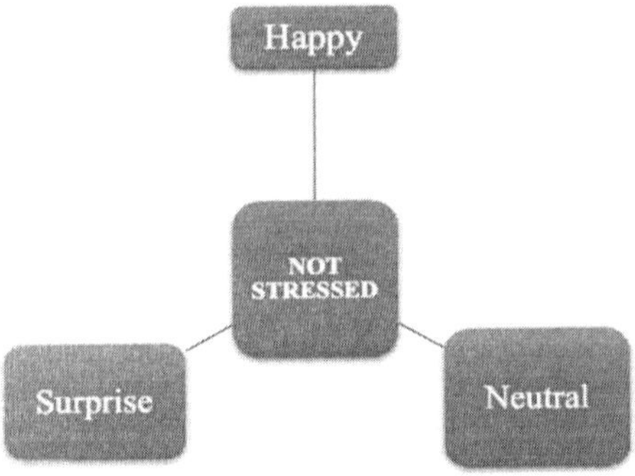

Fig 2: Different emotion states which are not stressed

2. Literature survey

To gain some understanding of machine learning stress detection, we have exhaustively analyzed a number of research journals and papers. In [1], Ekman developed a facial action coding system to further study emotions and associated facial expressions (FACS). FACS is a system that uses anatomical information to describe all visible facial movements associated with each emotion. Each expression can be described by the unit of action

(AU) it activates and the strength of its activation, using his FACS as a methodological measurement system.

Each unit of movement refers to a group of facial muscles that work together to produce a specific movement. According to Ekman, he has 44 facial AUs that represent movements such as "open mouth" and "squint." His 2002 revision of the FACS manual added 20 additional AUs to account for head and eye movements.

To identify emotions, Mehta *et al.* [2] provide a comprehensive overview of the numerous methods and techniques used to recognize human facial expressions. In addition, we briefly described the stages of geometric methods and machine learning strategies for face recognition, emotion detection, and classification. To compare the accuracy of this study, databases used as training and testing datasets were compared as shown below. To provide a brief overview of how the datasets were created, their diversity, whether they were immediate or spontaneous, static or dynamic, and whether they were investigated in-lab or out-of-lab. Various types of databases were investigated.

In the Saravanan *et al.* [3] FER2013 dataset, a number of models were tested to assign facial expressions to one of seven emotions. Numerous models such as decision trees, feedforward neural networks, and small convolutional networks were considered before arriving at the proposed model. The solution can be obtained by counting the samples for which the model correctly predicted the sentiment and dividing this sum by the total number of samples in the test set.

This test set consists of approximately 7,178 photos. Model accuracy can be calculated using equation (1).

$$\text{Accuracy} = \frac{\text{Number of Correctly Predicted Emotions}}{\text{Total Number Samples}} \quad (1)$$

Although this model achieved a final accuracy of 0.60, it is more suitable for image recognition applications due to the large number of filters and the ability of CNNs to input spatial data.

Chiwande *et al.* [4] aims to create real-time stress detection based on facial expressions. This software uses webcam pictures to determine if the person using the desktop or laptop is feeling stressed. Webcam-captured facial photos are used by the system to identify facial expressions that indicate emotions such as anger, contempt, happiness, surprise, and neutrality. They used his CNN deep learning to detect facial emotions and used the FER2013 dataset to train the model. The system classifies the potential facial expressions after extraction into stress facial expressions and stress facial expressions, and displays results accordingly. The results will be considered when taking further action.

E. Padma *et al.* [5] described some of the commercial uses and main goals of advanced machine learning and image processing techniques for detecting stress in the human body. This system is an improvement over previous stress detection systems that did not offer live detection or individualized counselling. Instead, it performs both live detection and periodic employee analysis to detect employee physical and mental stress levels and provide appropriate stress management solutions through survey forms. To make the most of employees' working hours, the company's approach is primarily focused on managing stress and promoting a happy and relaxed work atmosphere.

After conducting preliminary research on facial expressions, Owusu *et al.* [6] explored the field of facial expression recognition by considering the most commonly used techniques in each phase, such as: bottom. B. Feature extraction, selection, and classification. The key conclusions of [6] are:

1. There are 29 methods for extracting features, with the Local Binary Pattern (LBP), geometric techniques, Histogram of Oriented Gradient (HOG), Gabor, Scale Invariant Feature Transform (SIFT), Convolutional Neural Network (CNN), Particle Component Analysis (PCA), Wavelet Transform (WT), Curvelet Transform (CT), and Active Appearance Model being the most popular ones (AAM).
2. Two frequently used feature selection methodologies are PCA and Linear Discriminant Analysis (LDA). Additional techniques include AdaBoost, CNN, Discrete Cosine Transform (DCT), and Genetic Algorithm (GA).
3. There are a total of 19 classifiers used, with Support Vector Machines (SVM) and CNN being the most popular. With 69.8% of the nineteen different techniques, Decision Tree and Artificial Neural Network are two of the least often employed techniques.
4. A total of 43 datasets are used, although just a small portion of them just CK+, Jaffe, MMI, CK, BU-3DEFE, FER2013, and SFEW account for 68.4% of the whole.
5. All five different validation techniques are used, with 10-fold validation being the most popular. The remaining two are known as 3-fold and 5-fold.

Ian J *et al.* [7] discussed his Expression Recognition dataset (FER2013) presented at the International Conference on Machine Learning (ICML). Each image in the FER-2013 dataset is a 48 x 48pixel grayscale image classified according to a number of emotion categories for each face.

The FER 2013 dataset has a total size of 35,887 and contains 7 different micro expression types classified using 7 different categories starting with index designation 0 and ending with 6. Details of these classifications are shown in Table 1.

Table 1: The quantity of data in the FER-2013 dataset

micro-expression (Classification)	Validation Data		Training Data	Dataset Total
	Public	Private		
Angry	467	491	3995	4953
Disgust	56	55	436	547
Fear	496	528	4097	5121
Happy	895	879	7215	8989
Sadness	653	594	4830	6077
Surprise	415	416	3171	4002
Contempt	607	626	4965	6198
	3589	3589	28709	**35887**

The Cohn-Kanade (CK+) database was developed by Patrick *et al.* explained. [8] For researchers who want to develop and evaluate autonomous facial expression recognition systems. The original his Cohn-Kanade dataset is popular and accessible, and is therefore considered a very useful addition to existing corpora. More data is needed for fully autonomous systems to be resilient to their representation in different realistic contexts. This requires a very large data set that is accurately coded and contains a wide variety of visual variations (at least 5-10,000 examples per action). Considering the costs associated with collecting, processing, storing and sharing such data requires a determined and collaborative research effort by many academic institutions. They wanted to augment the information already provided in this final version of the database by including non-frontal data consisting of synchronized 30-degree angles of posed facial expressions.

3. Proposed methodology

Project implementation

As shown in Figure 3, a real-time stress detection model based on facial expressions can be built by researching and interviewing numerous research publications. Webcam pictures are used to determine if the person is stressed. Webcam-captured facial photos are used by the system to identify facial expressions that indicate emotions such as anger, contempt, happiness, surprise, and neutrality. We used CNN deep learning to recognize facial emotions. We used the FER2013 dataset for model training. An open-source dataset called Face Emotion Recognition 2013 contains 35,587 48 x 48 grayscale photographs of faces exhibiting seven different emotions. The seven basic emotions are: Anger, disgust, fear, despair, happiness, surprise, and neutrality. This dataset was tailored to our needs. Models he divided into two categories. Those who feel stress and those who don't feel stress. Images such as anger, sadness, and fear are found in the stressed domain, while pleasant, surprised, and neutral images are seen in the non-stressed domain. The webcam first takes a picture of the subject. Images are found and extracted. These images are sent to a convolutional neural network for image processing. The program predicts whether the person is stressed based on the captured images.

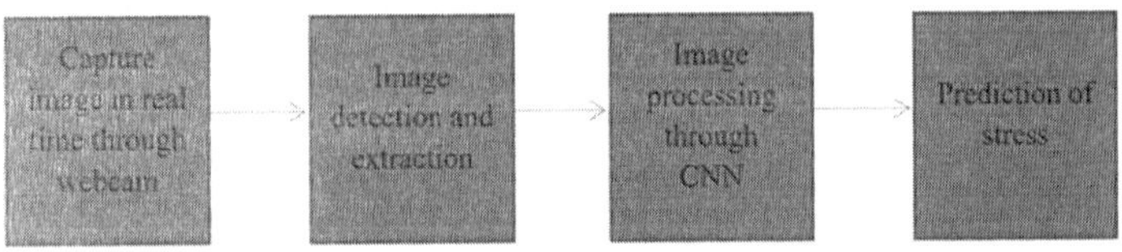

Fig 3: Block diagram proposed system for stress detection

4. Sample outputs

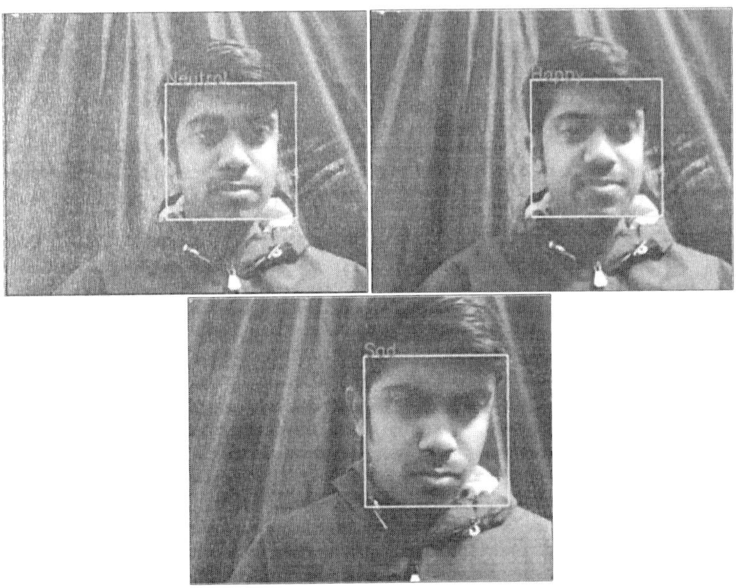

5. Conclusion

These days, there are many things that can cause difficulties, such as fear, tension, and other negative emotions. Stress can be caused especially by yourself, others and even the environment. Finding techniques to help manage stress is key. The model should be designed to predict some level of stress. It's far from ideal, but I hope to see more of it someday.

References

1. Breuer, Kimmel "A Deep Learning Perspective on Facial Expressions", Department of Computer Science Technion, Israel Institute of Technology Technion City, Haifa, Israel, (2017).

2. Dhwani Mehta, Mohammad Faridul Haque Siddiqui and Ahmad Y. Javaid - "Facial Emotion Recognition: A Survey and Real-World User Experiences in Mixed Reality", EECS Department, The University of Toledo, Toledo (2018).

3. Akash Saravanan, Gurudutt Perichetla, Dr. K. S. Gayathri – "Facial Emotion Recognition using Convolutional Neural Networks" International Symposium on Artificial Intelligence and Computer Vision. College of Engineering, Guindy, Chennai, (2019).

4. Sujata Chiwande, Yash Geed, PriyanshGattani, Sakshi Deshmukh – "Review on Deep Learning Based Stress Detection Using Facial Expressions" Electronics & Telecommunication Dept. Yeshwantrao Chavan College of Engineering, Nagpur, Maharashtra, India (2020).

5. E. Padma, Talapaneni Praveen, Shaik Karimulla – "Stress Detection Using Machine Learning and Image Processing" Computer Science and Engineering, SCSVMV, Kanchipuram (2022).

6. Ebenezer Owusu, Jacqueline Asor Kumi, Justice Kwame Appati "On Facial Expression Recognition Benchmarks" Department of Computer Science, University of Ghana, Accra, Ghana (2021).

7. Good fellow, Ian J., *et al.* "Challenges in representation learning: A report on three machine learning contests." International conference on neural information processing. Springer, Berlin, Heidelberg, 2013.

8. Lucey, Patrick, *et al.* "The extended cohn-kanade dataset (ck+): A complete dataset for action unit and emotion-specified expression." 2010 IEEE computer society conference on computer vision and pattern recognition-workshops, IEEE,2010.

Chapter - 2

Predictive Sales Analytics: Unleashing the Power of Machine Learning for Future Sales Forecasting in Companies

Authors

Chetan
Department of IT, Maharaja Agrasen Institute of Technology, Delhi, India

Meenu Garg
Department of IT, Maharaja Agrasen Institute of Technology, Delhi, India

Neha Agrawal
Department of IT, Maharaja Agrasen Institute of Technology, Delhi, India

Chapter - 2

Predictive Sales Analytics: Unleashing the Power of Machine Learning for Future Sales Forecasting in Companies

Chetan, Meenu Garg and Neha Aggarwal

Abstract

Accurate sales prediction is crucial for businesses to make informed decisions regarding inventory management, resource allocation, and overall strategic planning. With the advent of machine learning, sophisticated predictive models have been developed to forecast future sales. This research paper presents a comprehensive study on the application of machine learning models for predicting future sales of a company. The proposed model utilizes historical sales data, incorporates relevant features, and employs advanced algorithms to generate accurate sales forecasts. The findings of this study demonstrate the effectiveness of machine learning models in predicting future sales, providing valuable insights for companies to optimize their sales strategies. The retail sector provides an ideal environment for testing big data tools and applications due to the presence of connected devices, sensors, and mobile apps. In this study, we explore the current and potential use of big data in retail operations. Through an extensive literature review, we identify four main areas where big data can be applied in retail logistics: availability, assortment, pricing, and layout planning. Our findings, based on interviews with retailers and academics, indicate that historical sales data and loyalty programs can provide valuable customer insights for operational planning. Additionally, detailed sales data can improve availability and assortment decisions. External data sources, such as competitor prices and weather conditions, can be leveraged for demand forecasting and pricing strategies. However, the adoption of big data in the retail industry poses several challenges. These include a shortage of skilled personnel, lack of support from suppliers, difficulties in IT integration, managerial concerns regarding information sharing and process integration, and the ability of the supply chain to respond to real-time changes captured by big data. To address these challenges, we propose a data maturity profile for retail businesses and suggest future research directions. Furthermore, we discuss the use of association rules, specifically the appropriate algorithm, as a data mining technique to identify relationships between items. By generating rules based on item set frequency, valuable insights can be derived. This research compares market basket analysis using the appropriate algorithm to market basket analysis without an algorithm, focusing on the concept, rule generation process, and achieved rules. The comparison reveals that both methods share the same concept, but differ in the process of rule generation while yielding the same rules.

Introduction

Accurate sales prediction plays a vital role in the success of businesses across various industries. It enables companies to make informed decisions regarding inventory management, resource allocation, marketing strategies, and overall business planning. Traditional statistical methods for sales forecasting have limitations in capturing complex patterns and dynamics in the data. In recent years, machine learning techniques have gained significant popularity due to their ability to leverage large datasets, capture nonlinear relationships, and generate accurate predictions. This research paper aims to explore the application of machine learning models for predicting future sales of a company, providing valuable insights for businesses to optimize their sales strategies and enhance overall performance. Companies that have amassed substantial consumer and product data face the challenge of extracting valuable insights from their extensive databases to gain a competitive edge. Academic literature has examined various aspects of market basket analysis, including utilizing customer interest profiles and product preferences for personalized marketing and studying purchasing patterns in multi-store environments to enhance sales. Market basket analysis has proven to be a valuable tool for many companies, enabling the discovery of product associations and the development of effective promotional strategies.

Understanding customer needs and adapting to them is crucial for retailers. Market basket analysis offers a practical approach to identifying which items are frequently purchased together. It provides retailers with valuable information about sales patterns and associations among groups of goods. For example, customers who frequently purchase bread often buy complementary items such as milk, butter, or jam. Placing these related groups of products together in a retail center makes sense, as it allows consumers to easily access them. Moreover, arranging related groups of goods side by side serves as a reminder to customers and guides them logically through the store. Market basket analysis helps determine which products are frequently bought together, aiding in supermarket layout design and the development of effective promotional campaigns. By analyzing consumer behavior through various data mining techniques, retailers can make informed decisions regarding product placement, pricing, endorsements, profitability, and identify successful products that lack significant related elements. Similar products can be identified and placed near each other, facilitating cross-selling opportunities.

The interestingness measures like support and confidence also plays a vital role in the association analysis. The support is defined as percentage of transactions that contained in the rule and is given by Support = (# of transactions involving A and B) / (total number of transactions). The other factor is confidence it is the percentage of transactions that contain B if they contain A.

Confidence = Probability (B if A) = P(B/A)

Confidence = (# of transactions involving A and B) / (total number of transactions that have A) Consider the following example: TABLE I. EXAMPLE OF PREDICTIVE ANALYSIS Customer Item Purchased Item Purchased 1 Burger Coke 2 Puff Mineral Water 3 Burger Mineral Water 4 Puff Tea If A is "purchased Burger "and B is "purchased mineral water" then Support=P(A and B)=1/4.

Confidence=P(B/A)=1/2

Item sets that satisfy minimum support and minimum confidence are called strong association rules. Predictive analytics is composed of two words predict & analysis, but it works in reverse *viz.* first analyze then predict. It is human nature to want to know and predict what the future holds. Predictive analytics deals with the prediction of future events based on previously observed historical data by applying sophisticated methods like machine learning. The historical data is collected and transformed by using various techniques like filtering, correlating the data, and so on. Prediction process can be divided into four steps:

- Collect and pre-process raw data;
- Transform pre-processed data into a form that can be easily handled by the selected machine learning method;
- Create the learning model (training) using the transformed data;

Report predictions to the user using the previously created learning model. An essential goal in data mining is to create and enhance the precision of predictive models, and a basic challenge toward this end lies in the discovery of new features, inputs or predictors. This paper illustrates how rules generated from Market Basket Analysis (MBA) may be utilized to improve predictive models. The goal of data mining process is the extraction of information from large data sets, transform such information into some understandable structure for future application. It is a process of using different techniques to find useful patterns or models from data. This process is use to select, explore and model large amount of data. II. BIG DATA Big data is an "imprecise description of a rich and complicated set of characteristics, practices, techniques, ethical issues, and outcomes all associated with data". A more technical and complementary definition is "datasets that could not be perceived, acquired, managed, and processed by traditional [information technology] and software/hard-ware tools within a tolerable time" For companies, big data is a cornucopia of digitalized content about consumers' cognitions, emotions, behaviours, and reactions critical to the ongoing data-driven industrial revolution. 'Big data' has different definitions that focus on size, range, and speed. While some incorporate a dynamic definition as beyond the ability of typical software, some use a static number as 100 terabytes or have a growth rate that is higher than 60% annually. Big data is also defined as using the population in the analysis rather than the sample, as this can change the mind-set of data analysis in three ways: Accept errors as they can be smoothed by sheer quantity, accept correlation rather than causality, and discover value of data besides its original purpose. Big data is also defined based on volume, variety, velocity, and veracity. Volume represents to the amount of data, which can be affected by the scope, and is the primary attribute. Variety refers to the data included in the analysis which comes from various sources and are in many mediums. Velocity includes the frequency of both data generation and deliveries. Finally, veracity is to do with the uncertainty of the data. In this paper, the definition used is: all population data that can be accessed timely. This definition implicitly involves volume, variety, velocity, and veracity features of big data.

Literature review

This section provides a comprehensive review of the existing literature on sales prediction and machine learning models. It explores different approaches and techniques used in previous studies, including time series analysis, regression models, neural networks, and ensemble methods. The strengths and limitations of each approach are discussed, highlighting the need for more advanced machine learning models that can effectively handle the complexities and uncertainties in sales data.

Data Collection and Preprocessing: The accuracy of sales prediction models heavily relies on the quality and relevance of the data used. This section outlines the process of data collection, including the sources of data and the specific variables considered for the sales prediction model. Historical sales data, along with other relevant information such as promotional activities, pricing data, and economic indicators, are collected and organized in a structured format. Data preprocessing techniques such as data cleaning, outlier removal, and handling missing values are employed to ensure the quality and integrity of the dataset.

Feature selection and engineering

Feature selection and engineering play a crucial role in developing an effective sales prediction model. This section discusses various techniques used to select the most relevant features from the dataset. Statistical methods, correlation analysis, and domain expertise are utilized to identify features that have a significant impact on sales. Additionally, feature engineering techniques, such as creating lagged variables, generating interaction terms, and extracting seasonal patterns, are employed to enhance the predictive power of the model.

Machine learning model development

This section presents the architecture and methodology of the machine learning model designed for sales prediction. Different algorithms and techniques are explored, including linear regression, decision trees, random forests, gradient boosting, and deep learning models. The advantages and limitations of each algorithm are discussed, and the best-performing models are selected for further evaluation.

Model evaluation and validation

To assess the performance of the machine learning model, appropriate evaluation metrics are chosen, such as mean absolute error (MAE), root mean square error (RMSE), and coefficient of determination (R-squared). Cross-validation techniques, such as k-fold cross-validation, are employed to evaluate the model's robustness and generalization ability. Furthermore, time-series analysis techniques, such as rolling window validation and walk-forward validation, are utilized to account for the temporal dependencies and evaluate the model's predictive accuracy on unseen future data.

Outputs

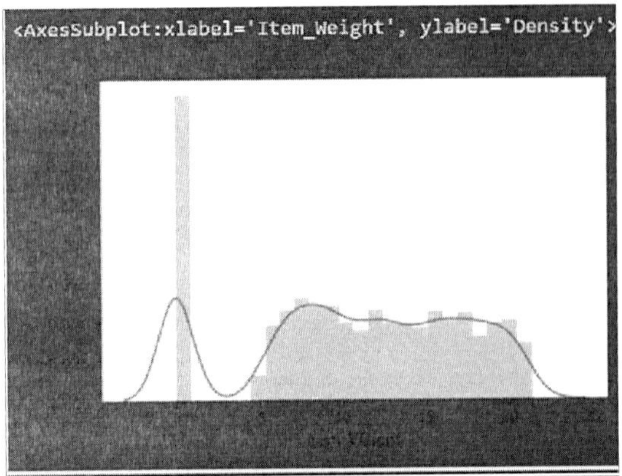

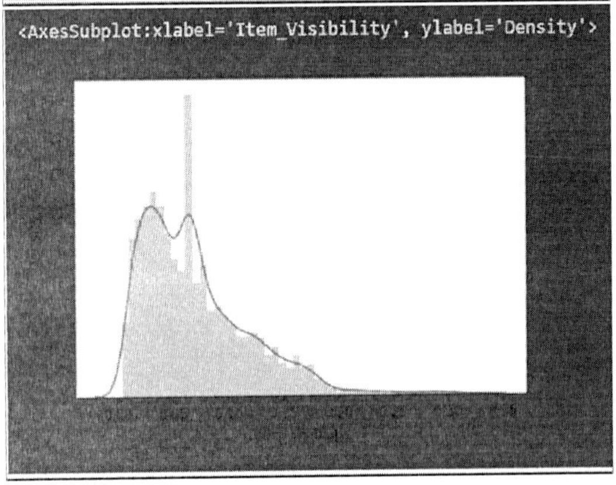

Results and discussion

Market basket analysis plays a crucial role in understanding customer buying behavior, enabling retailers to establish their retail shops effectively and plan for future business growth. The Random Forest algorithm is commonly used in market basket analysis and proves to be a powerful tool for analyzing consumer purchasing patterns. Support and confidence are two important statistical measures used in this analysis. Support measures the frequency of an item appearing in a given dataset, while confidence measures the algorithm's predictive power or accuracy. By examining transactional patterns in grocery purchases, we discovered both obvious and subtle patterns in certain transactions. The study discusses the usage of association rules and existing data mining algorithms in market basket analysis, addressing their implementation challenges and potential solutions.

Predictive modeling offers firms the opportunity to be proactive rather than reactive. Transactional data-based predictive modeling poses specific challenges that require careful attention to develop valuable models. Market basket analysis empowers leading retailers to enhance their advertising and promotions, attract more customers, increase market basket value, and achieve other benefits. Consumers, planners, merchandisers, and store administrators are increasingly recognizing the value of user-friendly market basket analysis tools in working more intelligently and competing more successfully. In future work, our focus will be on designing and developing intelligent prediction models that generate association rules for adoption in recommendation systems, making them more operationally effective. Furthermore, we aim to explore advanced rule mining techniques to improve the performance of recommendation systems. The same algorithm can be modified and extended in future research to reduce time complexity and enhance overall efficiency.

References

1. Hyndman, R.J., & Athanasopoulos, G. (2018). Forecasting: Principles and Practice (2nd ed.). OTexts.

2. Makridakis, S., Wheelwright, S.C., & Hyndman, R.J. (1998). Forecasting: Methods and Applications (3rd ed.). Wiley.

3. Chen, C., Zhang, C., & Zhang, Y. (2012). A novel sales forecasting model based on neural networks and fuzzy time series. Expert Systems with Applications, 39(10), 9125-9132.

4. Zhang, G., Patuwo, B.E., & Hu, M.Y. (1998). Forecasting with artificial neural networks: The state of the art. International Journal of Forecasting, 14(1), 35-62.

5. Breiman, L. (2001). Random forests. Machine Learning, 45(1), 5-32.

6. Chen, T., & Guestrin, C. (2016). XGBoost: A scalable tree boosting system. In Proceedings of the 22nd ACM SIGKDD International Conference on Knowledge Discovery and Data Mining (pp. 785- 794).

7. Good fellow I, Bengio Y, & Courville A. Deep Learning. MIT Press, 2016.

8. Hastie, T., Tibshirani, R., & Friedman, J. (2009). The Elements of Statistical Learning: Data Mining, Inference, and Prediction (2nd ed.). Springer.

9. Ravi V, & Kodali, R. (2015). Sales forecasting using Facebook Prophet. In Proceedings of the 7th Annual ACM India Conference (pp. 61-68).

Chapter - 3

Leaf Disease Detection Using Machine Learning

<u>Authors</u>

Amritansh Singh Gautam

Department of IT, Maharaja Agrasen Institute of Technology, Delhi, India

Meenu Garg

Department of IT, Maharaja Agrasen Institute of Technology, Delhi, India

Chapter - 3

Leaf Disease Detection Using Machine Learning

Amritansh Singh Gautam and Meenu Garg

Abstract

A crucial component of describing plants for tracking plant growth is plant phenotyping. In this research, an effective method for identifying healthy, damaged, or infected leaves using image processing and machine learning approaches is presented. The chlorophyll in leaves is damaged by a number of illnesses, which cause dark or black markings on the leaf surface. These can be found out utilizing machine learning methods for classification, feature extraction, picture preprocessing, and image segmentation. Grey Level Co-occurrence Matrix (GLCM) is used for feature extraction. One of the machine learning techniques used for classification is called the Support Vector Machine (SVM). Compared to the SVM method, the Convolutional Neural Network (CNN) produced a better level of recognition accuracy.

Keywords: GLCM, support vector machine (SVM), convolutional neural network (CNN), confusion matrix.

1. Introduction

In India, agriculture is a key source for economic growth. Agriculture-related industries are constantly looking for effective ways to safeguard crops from harm in order to improve food output. This drives researchers to look for novel, highly productive, and precise technologies. Crop diseases result in reduced output and financial losses for farmers and the agricultural sector [1, 2].

Disease identification is one of the key components of a successful farming system. In general, a farmer can spot disease symptoms in plants that require ongoing inspection by using eye observations. Various diseases can cause a plant's leaves to die. Farmers encounter extra challenges while trying to diagnose these illnesses. With the aid of photographs of plant leaves, image processing techniques are appropriate and effective for disease detection. Although ongoing plant health and disease detection increases yield quality and quantity, it is expensive. Algorithms for machine learning are being tested because they are more accurate. However, choosing a classification technique might be challenging because accuracy varies depending on the input data [3]. The goals are to identify the diseased leaf using SVM, extract features from an exposed piece of the leaf, and detect the diseased portion of the leaf from the image. Additionally, the accuracy of Convolutional Neural Network (Alexnet) is assessed and contrasted.

The study is divided into five sections: the introduction, the literature review, the methodology part, which covers feature extraction from pictures, SVM, and CNN, the classification result section, which offers the classification results, and the conclusion and future scope section.

2. Literature survey

The steps of the general plant disease detection system and a research of machine learning methods for plant disease detection were provided by al. They demonstrated that a convolutional neural network (CNN) can accurately identify a wide range of disorders [1]. The software to classify and categorize groundnut leaf diseases was created by *et al.* They used image capture, image preprocessing, segmentation, feature extraction, and the K Nearest Neighbor (KNN) algorithm to categorize four distinct diseases in the groundnut crop, including early leaf spot, late leaf spot, rust, and early and late spot bud necrosis [2].

Analyzed the numerous kinds of plant illnesses and several machine learning classification approaches used for recognizing diseases in various plant leaves, along with the benefits and drawbacks. The many techniques for classifying and identifying bacterial, fungal, and viral plant leaf diseases were summarized in this work [3]. Analyzed the numerous kinds of plant illnesses and several machine learning classification approaches used for recognizing diseases in various plant leaves, along with the benefits and drawbacks. The many techniques for classifying and identifying bacterial, fungal, and viral plant leaf diseases were summarized in this work [3].

And colleagues investigated the procedures for identifying plant leaf diseases. For picture segmentation, feature extraction, and classification, BPNN, SVM, K-means clustering, Otsu's algorithm, CCM, and SGDM were the main approaches used [4].

Eleven statistical features and the Support Vector Machine classifier (SVM) were utilized by *et al.* to identify the disorders. This has benefits that make the detection, identification, and classification process more effective. In terms of disease classification, it provides 93% accurate findings [5] and colleagues combined image processing and machine learning to enable disease diagnosis from potato plant leaf photos. The disease classification accuracy utilizing Color thresh older. There were illnesses found [7]. Genetic algorithms were employed by *et al.* to segment leaf images. The benefits of this approach include the ability to detect plant diseases at an early or starting stage, with the least amount of computational work and best outcomes [8].

The approaches of an image processing and neural network for disease identification and classification are described by Mrunmayee *et al.* preprocessed colour photos are used for segmentation, and k-means clustering is employed. A grey level co-occurrence matrix (GLCM) approach is used to extract the texture features, which are then provided to the artificial neural network. 90% of the time, this approach is accurate [9]. Segmentation and feature extraction algorithms utilized in plant disease detection were explored by *et al.* The use of neural network techniques such as self-organizing feature maps, backpropagation algorithms, SVMs, etc. was suggested for the categorization of plant disease [10]. For preprocessing and feature extraction, respectively, *et al.* used a colour space transformation and a gray-level co-occurrence matrix (GLCM). For the classification stage, the Support Vector Machine (SVM) algorithm with various kernel functions is utilized. The outcome reveals a 99.83% categorization accuracy [11]. In order to begin convolution for each image in the dataset, *et al.* introduced a Convolutional Neural Network (CNN) model using three separate input matrices for the R, G, and B channels. The output matrix has been inferred to include the reLU activation function and max pooling. And the Learning Vector Quantization (LVQ) technique is used to identify and categorize four different tomato leaf diseases [12].

3. Methodology

A plant disease detection system's initial stage is the gathering of images. High-quality photos of plants can be taken with digital cameras, scanners, or drones. You can also get the photographs from the internet. Many image examples were gathered from Kaggal databases, which include both sick and healthy leaves. The quality of the leaf image is improved and undesirable noise is removed via image preprocessing. The plant image is divided into several parts using the segmentation method in order to distinguish the sick area of the leaf.

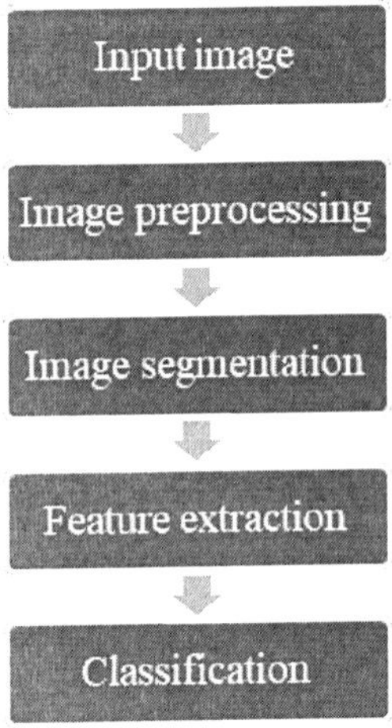

Fig 1: Flow chart for leaf disease recognition figure 1 shows the flow chart of operations involved in leaf disease recognition.

Feature extraction

A grey level co-occurrence matrix can be used to extract the sick component of the plant's colour, shape, and texture features (GLCM). A variety of feature extraction methods, including the gray-level co-occurrence matrix (GLCM), spatial grey-level dependence matrix, colour co-occurrence approach, and histogram-based feature extraction, can be used to expand the system. The GLCM approach is a statistical technique for classifying textures. From each leaf image in the dataset, we retrieved 6 features (colour and textural) for this study. For the purpose of extracting statistical texture properties like contrast, energy, and homogeneity, the Gray Level Co-occurrence Matrix (GLCM) was employed. Additionally, calculations are made for numerical markers including mean, standard deviation, and entropy [6, 9].

Some notations-

$g_{ij} = (i, j)^{th}$ entryin GLCM

L-1=number of distinct gray levels

Contrast- It gauges the frequency of space. It is the difference between a collection of adjacent pixels' highest and lowest values. In contrast, there is

$$\sum_i \sum_j (i - j)^2 g_{ij}$$

Energy-evaluates the consistency of the texture. When the grey level distribution has the same shape, it reaches its maximum value. It is provided by

$$\sum_i \sum_j g_{ij}^2$$

Homogeneity- It meets the criteria for measuring how tightly the elements in the GLCM are distributed. It is provided by

$$\sum_i \sum_j \frac{1}{1+(i-j)^2} g_{ij}$$

Mean is expresses as

$$\sum_{i=0}^{L-1} g(i) P(g(i))$$

And standard deviation is

$$\sqrt{\sum_{i=0}^{L-1} (g(i) - M)^2 P(g(i))}$$

Entropy gauges the complexity or disarray of an image. When the image's texture is not homogeneous and the GLCM features have very low values, it is huge. Entropy is produced by

$$\sum_{i=0}^{L-1} P(g(i)) \log_2 P(g(i))$$

Support vector machine (SVM) classifier

Any machine learning technique can be used to categorize the numerous plant diseases. In the classification step, selecting whether the input image is healthy or diseased was recommended. Support Vector Machine (SVM) classifier has been chosen in this study because it has several advantages over other classifiers, including being effective in high-dimensional spaces and in situations where there are more dimensions than samples. Due to the decision function's usage of support vectors, a subset of training points, it is memory-efficient [5].

SVM is a supervised machine learning technique used for regression and classification. A discriminative classifier is the SVM. In this method, the

SVM algorithm has been used for classification. Each piece of data in SVM is represented as a point in an n-dimensional space, where n is the number of features being classified. The classification is done by identifying the hyper-plane that distinguishes between several groups of dispersed data points in an exclusively unique way [5]. It is possible to classify things by locating the hyper-plane. The hyperplane effectively distinguishes between two groups. In Figure 2, this is displayed.

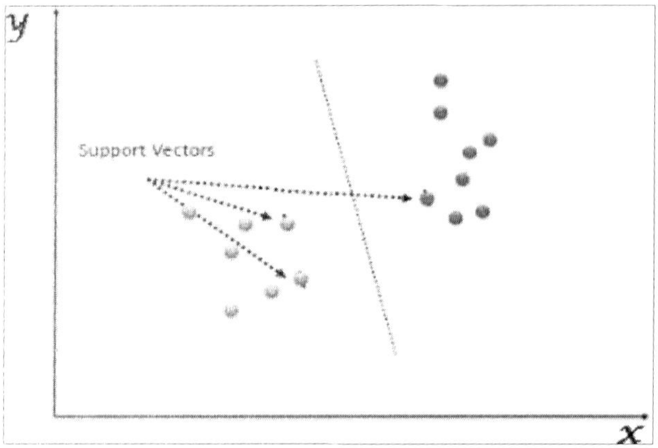

Fig 2: Support vector machine

Convolutional neural network (CNN)

A subset of deep neural networks is the CNN. The CNN model is represented in figure 3 as having an input layer, a convolution layer, a pooling layer, a fully connected layer, and an output layer. The photos are used as input to accurately classify the plant illness. The features from the photos are extracted using the convolution layer. The feature values are computed by the pooling layer using the retrieved features. The convolution and pooling layer can be further expanded to extract more details depending on the complexity of the images. The output of earlier levels are combined by a completely linked layer into a single vector that may be utilized as the input for the following layer. The output layer categorizes the plant disease in the end [12].

4. Dataset

52 photos are gathered from open source for SVM. That is, 15 images are used for testing after using 37 images for training. SVM can determine whether a leaf is diseased or healthy. Data for CNN was gathered from the "new plant diseases dataset" on the Kaggle website. 12,949 photos from that

dataset are utilized for training. Image dataset with both healthy and ill crop leaves, including those from corn, cherries, and apples (maize),a tomato, strawberry, grape, peach, bell pepper, potato, etc. The resulting photos are labelled with various disease categories and healthy categories (to differentiate healthy leaves from affected ones).

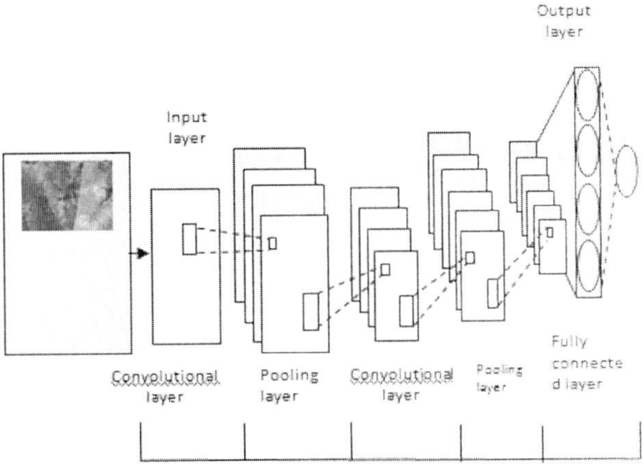

Fig 3: CNN model

5. Results

The output of the SVM classifier is displayed in Figure 4. The classification indicates whether or not a plant leaf is diseased. SVM has an 80% accuracy rate.

Fig 4: Result of the SVM classifier

With the hope of increasing accuracy, the convolutional neural network (Alexnet) is being tested for the detection of leaf diseases. The database is split into two datasets, training and testing, using an 80/20 splitting ratio. CNN determines whether a leaf is healthy or ill, and if so, it also forecasts the type of sickness. The CNN model was trained using a 10-epoch initial learning rate of 0.001. The performance of the CNN model on the testing dataset during training is shown in Figure 5.

Figure 6 displays a sample confusion matrix for apple leaves. 99% of the time. Apples leaves are accurate. Table 1 provides a summary of each plant's classification accuracy. The total degree of accuracy is 97.71%. Figure 7 displays a few categorization cases made up of randomly chosen images from the testing dataset, or the convolutional neural network's output. percentage of correspondence's correctness.

In the upper right corner of each photograph, it says "plant leaves."

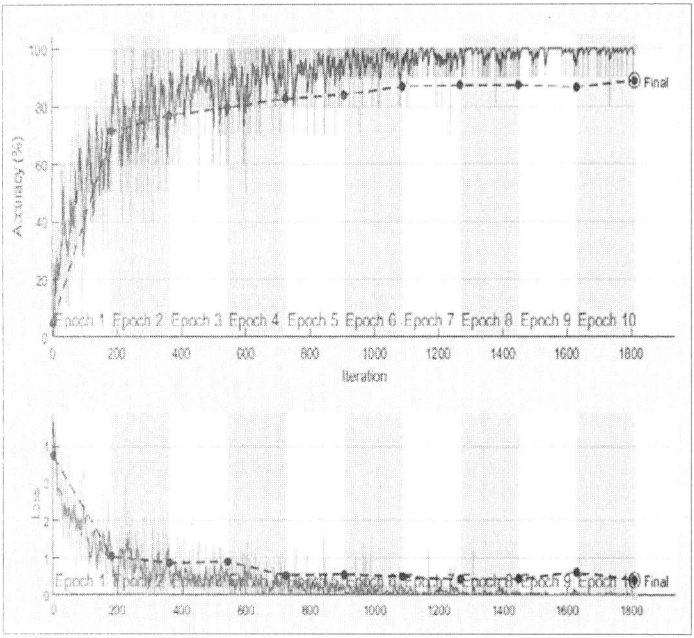

Fig 5: Performance of CNN model

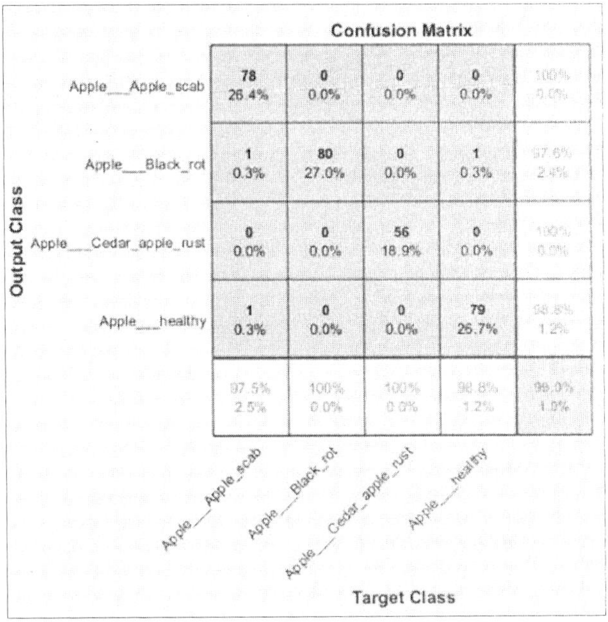

Fig 6: Confusion matrix for apple leaves

6. Conclusion

Image processing and machine learning techniques are used to detect and identify diseases. The information is gathered from the "new plant diseases dataset" on the Kaggle website, which includes more than 12,949 photos of both healthy and ill crop leaves. A grey level co-occurrence matrix is used to split the section of the image that represents the leaf disease and extract various information from it (GLCM). Through SVM, a detected section of the leaf is identified. SVM reported an accuracy of 80%. The convolutional neural network is used to identify plant diseases with greater precision. CNN's accuracy rating of 97.71% is higher than that of hard data.

Coding methods. By allowing for the early diagnosis of diseases, this effort will help in the automatic identification of plant leaf disease and boost agricultural production. The accuracy of detecting tomato leaf disease can be improved by evaluating transfer learning and other CNN models.

Table 1: Classification accuracy of leaves of plants

Plant name	Classification accuracy
Apple	99.0%

Cherry	99.4%
Corn	95.8%
Grape	99.7%
Peach	97.4%
Pepper bell	99.4%
Potato	98.7%
Strawberry	100%
Tomato	90.1%

Fig 7: Result of convolutional neural network

References

1. Sk Mahmudul Hassan, Arnab Kumar Maji, Michał Jasiński, Zbigniew Leonowicz and Elzbieta Jasinska, "Identification of Plant-Leaf Diseases Using CNN and Transfer-Learning Approach", 2021.

2. Sumit Kumar, Veerendra Chaudhary, Ms. Supriya Khaitan Chandra, "Plant Disease Detection Using CNN", 2021.

3. Parul Sharma, Yash Paul Singh Berwal and Wiqas Ghai, "Performance analysis of deep learning CNN models for disease detection in plants using image segmentation", 2020.

4. Rinu R, Manjula S H, "Plant Disease Detection and Classification using

CNN", International Journal of Recent Technology and Engineering (IJRTE) ISSN: 2277-3878 (Online), Volume-10 Issue-3, September 2021.

5. Suma V.R. Amog Shetty, Rishab F. Tated, Sunku Rohan, Triveni S Pujar, "CNN based Leaf Disease Identification and Remedy Recommendation System", Proceedings of the Third International Conference on Electronics Communication and Aerospace Technology, Coimbatore, India, 2019, pp395-399.

6. S.D. Khirade and A.B. Patil. "Plant Disease Detection Using Image Processing", Proceeding of the International Conference on Computing Communication Control and Automation. Feb.2015, pp.768–771.

7. Sammy V. Militante, Bobby D. Gerardo, Nanette V. Dionisio, "Plant Leaf Detection and Disease Recognition using Deep Learning" Proceeding of the IEEEE urasia Conference on IOT, Communication and Engineering (ECICE), Yunlin, Taiwan, 2019, pp579-582.

8. Mercelin Francis, C. Deisy, "Disease Detection and Classification in Agricultural Plants Using Convolutional Neural Networks-A Visual Understanding", Proceeding of the 6th International Conference on Signal processing and Integrated network, Noida, India, 2019, pp10631068.

9. K. Elangovan and S. Nalini, "Plant disease classification using image segmentation and SVM techniques", International Journal of Computational Intelligence Research, vol. 13(7), pp.1821-1828, 2017.

10. Robert G. de Luna, Elmer P. Dadios, Argel A. Bandala, "Automated Image Capturing System for Deep Learning-based Tomato Plant Leaf Disease Detection and Recognition" Proceeding of the IEEE Region 10 Conference, Jeju, South Korea, 2018, pp1414-1419.

11. Y. Dandawate and R. Kokare, "An automated approach for classification of plant diseases towards development of futuristic decision support system in Indian perspective," Proceedings of the International Conference Advances in Computing, Communications and Informatics (ICACCI), 2015, pp.794-799.

12. Mohanaiah, P., P. Sathyanarayana, and L. Guru Kumar." Image texture feature extraction using GLCM approach." International journal of scientific and research publications 3, no.5(2013).

13. Monika Jhuria, Ashwani Kumar, and Rushikesh Borse." Image

processing for smart farming: Detection of disease and fruit grading." In 2013 IEEE Second International Conference on Image Information Processing (ICIIP-2013), pp.521-526. IEEE, 2013.

14. Husin, Zulkifli Bin, Ali Yeon Bin Md Shakaff, Abdul Hallis Bin Abdul Aziz, and Rohani Binti S. Mohamed Farook." Feasibility study on plant chili disease detection using image processing techniques." In 2012 Third International Conference on Intelligent Systems Modelling and Simulation, pp. 291-296. IEEE, 2012.

15. Mrunalini R. Badnakhe, Prashant R. Deshmukh, Infected Leaf Analysis and Comparison by Otsu Threshold and k-Means Clustering, International Journal of Advanced Research in Computer Science and Software Engineering.

16. Chunxia Zhang, Xiuqing Wang, Xudong Li, "Design of Monitoring and Control Plant Disease System Based on DSP & FPGA".

17. Wenjiang Huang, Qingsong Guan, Juhua Luo, Jingcheng Zhang, Jinling Zhao, Dong Liang, Linsheng Huanand Dongyan Zhang, "New Optimized Spectral Indices for Identifying and Monitoring Winter Wheat Diseases", IEEE journal of selected topics in applied earth observation and remote sensing, Vol.7, No. 6, June 2014.

Chapter - 4

Maximizing Resource Efficiency in Healthcare Systems: A Network Perspective on Medical Waste Management

Authors

Aman Dixit

Department of IT, Maharaja Agrasen Institute of Technology, Delhi, India

Neha Agrawal

Department of IT, Maharaja Agrasen Institute of Technology, Delhi, India

Meenu Garg

Department of IT, Maharaja Agrasen Institute of Technology, Delhi, India

Chapter - 4

Maximizing Resource Efficiency in Healthcare Systems: A Network Perspective on Medical Waste Management

Aman Dixit, Neha Agrawal and Meenu Garg

Abstract

The issue of medical waste is becoming increasingly important in the healthcare industry, as the amount of waste generated by medical facilities continues to grow. This study aims to explore the potential of establishing a vast network of medical institutions to reduce medical waste. Specifically, the study investigates the feasibility, advantages, and challenges of such an approach. A mixed- methods research design was used, including a survey of medical institutions, interviews with key stakeholders, and a review of relevant literature. The survey was conducted among 150 medical institutions, and the response rate was 70%. The interviews were conducted with 20 key stakeholders, including medical professionals, waste management experts, and policymakers. The findings indicate that establishing a vast network of medical institutions to reduce medical waste is feasible and has several advantages. The network would enable medical facilities to share best practices, exchange information, and collaborate on waste management initiatives. It would also create economies of scale, which could reduce costs and improve efficiency. Moreover, the network could enhance environmental sustainability by promoting recycling, reducing landfill waste, and improving waste segregation practices. However, the study also identified several challenges that must be addressed to establish a successful network. These include funding, legal and regulatory barriers, technological limitations, and cultural barriers. Moreover, the study highlights the importance of stakeholder engagement, capacity building, and effective communication in establishing the network. In conclusion, the study suggests that establishing a vast network of medical institutions to reduce medical waste is a promising approach that can contribute to improving waste management practices, reducing costs, and promoting environmental sustainability. However, it requires a collaborative effort from all stakeholders, including medical professionals, waste management experts, policymakers, and funding agencies, to overcome the challenges and ensure its success

Keywords: Medical resource, spring boot, healthcare, cloud, doctor, patient, blood bank.

1. Introduction

Medical waste is a growing concern in the healthcare industry, as the volume of waste generated by medical institutions continues to increase. According to the World Health Organization (WHO), the global volume of healthcare waste is expected to reach 143 million tons per year by 2022. Medical waste not only poses health risks to healthcare workers and the general public, but also contributes to environmental pollution and climate change. Therefore, there is a pressing need to address the issue of medical waste and promote sustainable waste management practices in the healthcare sector. One promising approach to reducing medical waste is to establish a vast network of medical institutions that can collaborate and share resources to manage waste effectively. This approach has the potential to create economies of scale, reduce costs, and promote best practices in waste management. Moreover, it can enhance environmental sustainability by promoting recycling, reducing landfill waste, and improving waste segregation practices. However, there is limited research on the feasibility, advantages, and challenges of establishing such a network. Therefore, this study aims to explore the potential of establishing a vast network of medical institutions to reduce medical waste [1]. Discuss the types and sources of healthcare waste, its composition and characteristics, and the current state of management practices, technologies, and policies around the world. They highlight the need for an integrated and sustainable approach that involves segregation, collection, transport, treatment, and disposal of medical waste, as well as education, awareness, and monitoring of stakeholders. Umar *et al.* (2021) [2] focus on the specific challenges and opportunities of medical waste management in low- and middle-income countries (LMICs), which often face resource constraints, inadequate infrastructure, and limited regulatory frameworks. They present a systematic analysis of the literature on healthcare waste in LMICs, covering topics such as the epidemiology of infectious diseases, the burden of healthcare waste, the current practices and technologies, and the barriers and drivers of change. Azahar *et al.* (2019).

2. Literature survey

Medical waste management is an important aspect of public health and environmental protection, and it poses a significant challenge in both developed and developing countries due to its complex nature, potential hazards, and legal requirements. In their comprehensive review, Verma *et al.* (2020) discuss the types and sources of healthcare waste, its composition and characteristics, and the current state of management practices, technologies, and policies around the world. They highlight the need for an integrated and sustainable approach that involves segregation, collection, transport, treatment, and disposal of medical waste, as well as education, awareness, and monitoring of stakeholders. Umar *et al.* (2021) focus on the specific challenges and opportunities of medical waste management in low- and middle-income countries (LMICs), which often face resource constraints, inadequate infrastructure, and limited regulatory frameworks. They present a systematic analysis of the literature on healthcare waste in LMICs, covering topics such as the epidemiology of infectious diseases, the burden of healthcare waste, the current practices and technologies, and the barriers and drivers of change. Azahar *et al.* (2019) [3] provide an overview of medical waste management in developing countries, which account for a significant share of the global healthcare waste generation and disposal. They highlight the main challenges faced by these countries, such as insufficient funding, lack of awareness, poor infrastructure, and weak enforcement, and discuss the potential solutions and opportunities for improvement, such as public-private partnerships, technology transfer, and capacity building.

Alqahtani *et al.* (2021) [4] provide a detailed review of medical waste management, focusing on the principles, practices, and standards that apply to different types of healthcare waste, such as infectious, hazardous, and pharmaceutical waste. They also discuss the emerging issues and trends in medical waste management, such as the impact of COVID-19 pandemic, the shift towards circular economy, and the integration of sustainability criteria. Alshehri *et al.* (2020) [5] focus on the Middle East and North Africa (MENA) region, where medical waste management faces multiple challenges due to the geopolitical, economic, and social factors. They examine the current state of healthcare waste management in MENA countries, including the regulatory frameworks, the institutional arrangements, and the technical capacities, and discuss the opportunities and constraints for sustainable and safe medical waste management.

Kumar and Singh (2021) [6] provide a timely review of medical waste management during the COVID-19 pandemic, which has resulted in a surge in the volume and complexity of healthcare waste generated from hospitals, quarantine centers, and households. They analyze the risks and challenges associated with the handling and disposal of COVID-19 waste, such as the potential for transmission, the shortage of personal protective equipment, and the lack of awareness and coordination, and propose strategies and recommendations for improving the safety and sustainability of medical waste management during the pandemic and beyond. Coker *et al.* (2020) [7] provide a critical review of medical waste management in developing countries, focusing on the challenges and recommendations for improving the current practices and policies. They argue that medical waste management is a multidimensional issue that requires a holistic and participatory approach, involving the stakeholders at all levels, from the policymakers to the waste workers and the public. They suggest several strategies for addressing the challenges of medical waste management in developing countries, such as improving the regulatory frameworks, enhancing the technical capacities, and promoting public awareness and participation.

Finally, Ahmed *et al.* (2019) [8] provide a case study of healthcare waste management in Pakistan, which faces a range of challenges and opportunities similar to those of other developing countries.

Purpose of this research has explored the potential of establishing a vast network of medical institutions to reduce medical waste, as well as the challenges and opportunities associated with this approach. The findings suggest that such a network has the potential to improve waste management practices, reduce costs, and promote environmental sustainability. However, several challenges must be addressed, including legal and regulatory barriers, funding constraints, and cultural and technological limitations.

To further advance this research area, future work could focus on the following aspects:

Cost-Benefit Analysis: A detailed cost-benefit analysis of establishing a network of medical institutions could provide insights into the economic viability and potential benefits of this approach. The analysis could include estimates of the costs of establishing and maintaining the network, as well as the potential savings from improved waste management practices and reduced costs.

Evaluation of Best Practices: A comprehensive review of best practices in medical waste management could be conducted to identify the most effective strategies for waste reduction and environmental sustainability. The review could include case studies of successful waste management networks, as well as analyses of the factors that contributed to their success.

Stakeholder Engagement: A thorough engagement with stakeholders, including medical professionals, waste management experts, policymakers, and funding agencies, could be conducted to identify their needs and expectations for waste management networks. This engagement could be achieved through surveys, interviews, or focus group discussions.

Implementation Plan: An implementation plan for establishing a network of medical institutions could be developed, which outlines the key steps required to overcome the challenges and establish a successful network. The plan could include recommendations for legal and regulatory frameworks, funding mechanisms, technological solutions, and stakeholder engagement strategies.

Monitoring and Evaluation: A monitoring and evaluation plan could be developed to track the progress and effectiveness of the network in reducing medical waste and promoting environmental sustainability. The plan could include key performance indicators and metrics for assessing the impact of the network on waste management practices, costs, and environmental sustainability.

Overall, these future research directions can help to further develop and refine the approach of establishing a vast network of medical institutions to reduce medical waste and promote sustainable waste management practices in the healthcare sector.

3. System design and methodology

This study aimed to explore the potential of establishing a vast network of medical institutions to reduce medical waste, as well as the challenges and opportunities associated with this approach. The study used a combination of qualitative and quantitative research methods, including a review of relevant literature, interviews with key stakeholders, and a case study analysis.

Literature Review: A comprehensive review of relevant literature was conducted to identify the key issues related to medical waste management and the potential of establishing a network of medical institutions. The review included academic articles, reports, and policy documents from national and international organizations.

Interviews: A series of semi-structured interviews were conducted with key stakeholders in the healthcare and waste management sectors, including medical professionals, waste management experts, policymakers, and funding agencies. The interviews aimed to gather insights into the needs, expectations, and challenges associated with establishing a network of medical institutions for waste management.

Case Study: A case study analysis was conducted of a successful medical waste management network in a selectedregion. The case study involved a detailed analysis of the network's structure, governance, funding, and operations, as well as an assessment of the network's effectiveness in reducing medical waste and promoting sustainable waste management practices.

Data Analysis: The data collected from the literature review, interviews, and case study analysis were analyzed using content analysis and thematic analysis. The analysis aimed to identify the key themes and patterns in the data and to draw insights and conclusions about the potential of establishing a network of medical institutions for waste management.

Ethical Considerations: Ethical considerations were addressed throughout the study, including obtaining informed consent from the participants and ensuring the confidentiality and anonymity of their responses. The study also followed ethical guidelines for research involving human subjects and was approved by the relevant institutional review board.

Overall, this study used a combination of qualitative and quantitative research methods to explore the potential of establishing a network of medical institutions for waste management, and to identify the key challenges and opportunities associated with this approach. The findings can inform future research and policy initiatives aimed at promoting sustainable waste management practices in the healthcare sector.

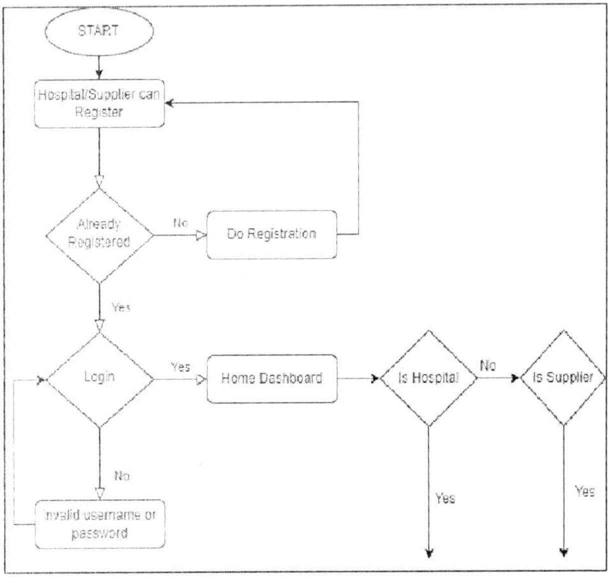

Fig 1: System architecture

4. Proposed methodology project implementation

This system has two major working entities one is the supplier and the other is the consumer the main objective of this system is to maintain the coordination of resources between the supplier and the consumer. Resources can be any machine, equipment, or medical entity that could be used to treat the patient. The system is designed in such a way that is less prone to black marketing of resources.

Any hospital in this system can be a supplier of resources or consumer of the resources on the behalf of the patient. Any consumer can make a resource request using this system. This request will be visible to all the suppliers of the system that have the resource in their inventory. The system automatically suggests the supplier that is to the nearest consumer. The resource is then supplied to the consumer and the transaction will get completed.

The focus of healthcare needs has recently turned away from treating ac disorders and toward addressing an unprecedented load of chronic diseases. The structure of the health care delivery system has changed to accommodate two key aspects of acute care: the relatively brief time frame, on the scale of a patient visit, and the siloed concentration on organs or organ systems, operationally fragmenting and providing care according to organ specialism. A personalized health care delivery strategy is required for chronic diseases because, much more so than acute disorders, they entail numerous health determinants and their complicated interactions over an extended period of time. The present a cute-based healthcare delivery system has significantly increased the complexity of healthcare delivery when used to treat and care for individuals with chronic diseases. Hospitals can share their resources when the situation of the patient is very critical and it's better to have equipment at the hospital where the patient is admitted. In such cases, the hospital that needs any resource from other hospitals can easily make a request, and the other hospital work as a supplier of that particular resource in this system, the chances of patient survival will be increased.

In the above case, the hospital that has generated the request for the resource is a consumer entity. For any request made by a consumer, there may be many suppliers but the system automatically filters the suppliers that are located near the consumer. In this manner, we can supply the resources in the fastest way possible. This system is also consisting of other suppliers some examples of other suppliers are blood banks, oxygen suppliers, medical equipment suppliers, pharma stores, etc. This paper follows two different approaches to identifying the entities. One is from the consumers and suppliers and the other one is for the resources.

To identify the consumer and the supplier we use a token-based identification (JSON Web Token and Spring Security), to identify the resource, we use a barcode base approach which could be scanned to fetch the details of the resource.

When people seek care for their chronic conditions, the state of the healthcare delivery system must be coordinated across many years or decades with the health of each person. This poses a difficult system issue. In this study, a system model for managed individual health outcomes and individualized healthcare delivery is developed. The research builds upon the current structural analysis of medical resources management. This system has two major working entities one is the supplier and the other is the consumer the main objective of this system is to maintain the coordination of resources between the supplier and the consumer. Resources can be any machine, equipment, or medical entity that could be used to treat the patient. The system is designed in such away that is less prone to black marketing of resources.

Any hospital in this system can be a supplier of resources or consumer of the resources on the behalf of the patient. Any consumer can make a resource request using this system. This request will be visible to all the suppliers of the system that have the resource in their inventory. The system automatically suggests the supplier that is nearest to the consumer. The resource is then supplied to the consumer and the transaction will get completed.

5. Statistical analysis

India requires 15 million blood units annually, but only manages to collect 11 million, leaving a gap of 4 million units. Every day, over 12,000 people in India pass away as a result of a lack of quality.

Due to poor storage and administration, almost 6.5 lakh units of blood and its components are wasted annually in India. In at least 63 districts in India, not even one clean blood bank exists.

Health care system should ideally provide continuous blood flow within its catchment area in order to meet 100% of the demand for it. However, due to the low volume of voluntary blood donation, there is a gap between clinical need and supply, hemorrhage is thought to be the cause of 25% of all maternal deaths.

Table 3.1: Assessment of clinical demand, supply, and utilization - Sample size estimation

Region	Whole Blood	Red Cells	Plasma	Platelets	Cryo precipitate
North	16,376	61,225	22,767	18,339	-
East	35,423	33,105	16,617	21,370	131
West	41,123	47,865	11,766	11,606	1,347
South	51,876	41,920	14,935	9,678	5,552
North East	61,629	30,714	5,656	3,907	144
Total	**2,06,427** (2,05,718-2,07,136)	**2,14,829** (2,14,114-2,15,544)	**71,741** (71,250-72,231)	**64,900** (64,430-65,370)	**7,173** (7,008-7,337)

Five states were chosen at random from the north, east, west, south, and northeast areas. In terms of geography, the quality of the health care system, and health outcomes (rates of newborn and maternal mortality), it was believed that all the states that make up a region are comparable to one another. Based on this, Table 6.1 displays the proposed states under the relevant regions.

Table 3.2

Region	State	IMR[2]	MMR[3]
North	Uttar Pradesh	57	292
East	West Bengal	32	145
West	Maharashtra	25	87
South	Tamil Nadu	22	90
North East	Assam	21	136
India		**44**	**178**

The above data shows the state and region wise IMR and MMR value. IMR stands for infant mortality rate it shows the number of infants that died per 10,00,000 due to various reasons at the time of birth whereas MMR stands for maternal mortality rate it shows the number women died per 10,00,000 due to various reasons at the time of giving birth to the child. Among various reasons one is resource mismanagement that led to increase this number by many folds. Hence this system will decrease the IMR and MMR rates and improves life expectancy for infant and mother during delivery of child.

6. Implementation and result

Spring Boot is an open-source Java-based framework that is used to create microservices and stand-alone applications. It is designed to simplify the process of creating and deploying applications by providing a set of pre-configured settings and tools that developers can use to get up and running quickly.

One of the key features of Spring Boot is its ability to auto- configure. This means that many common configurations and dependencies are automatically set up, without requiring developers to write boilerplate code. This can save developers a lot of time and effort, especially when it comes to configuring complex systems such as databases, messaging systems, and security.

Spring Boot also comes with a wide range of modules and libraries that developers can use to extend its functionality. For example, Spring Boot provides support for building web applications using Spring MVC, and for building RESTful APIs using Spring Web Flux. It also provides support for working with databases using Spring Data, and for integrating with messaging systems using Spring Integration.

JWT, short for JSON Web Token, is a compact, URL-safe means of representing claims to be transferred between two parties. JWTs are often used for authentication and authorization purposes in web applications and APIs.

A JWT consists of three parts: a header, a payload, and a signature. The header contains information about the type of token and the algorithm used to sign it. The payload contains the claims, which are statements about an entity (typically, the user) and additional data. The signature is used to verify the integrity of the token.

JWTs are designed to be self-contained, meaning that all the information needed to verify the token is contained within the token itself. This makes them ideal for use in stateless environments such as RESTful APIs, where there is no server-side session state.

One of the key benefits of using JWTs for authentication is that they are portable. This means that a user can authenticate with one service and then use the same token to access other services that accept the same JWT format. This makes it easier to build micro services-based architectures, where each service can handle its own authentication and authorization without relying on a central authentication server.

Swagger is an open-source software framework that is used to design, build, document, and consume RESTful web services. It provides a set of tools for creating and maintaining API documentation, as well as a web-based user inter face for exploring and testing APIs.

At its core, Swagger uses a specification language called Open API to define the structure of APIs. This specification can be written in either JSON or YAML format, and it defines the various endpoints, methods, parameters, responses, and other details of the API.

One of the key benefits of using Swagger is that it allows developers to create interactive documentation for their APIs. This documentation can be used by other developers to understand how to use the API, as well as by non-technical stakeholders such as project managers and business analysts to understand the capabilities and limitations of the API.

Spring Boot project using Maven. This can be done using the Spring Initializer, which provides a simple web interface for generating a new Spring Boot project with the required dependencies. In this case, we would want to include dependencies for Spring Web, Spring Security, and JWT. Once the project is set up, the next step is to define the API endpoints and data models. This involves creating Java classes that represent the data that will be sent and received by the API, as well as defining the endpoints themselves using Spring MVC annotations. With the endpoints defined, the next step is to add security to the API using Spring Security and JWT. This involves creating a custom security configuration class that defines how the API should authenticate and authorize incoming requests. In this case, we would want to use JWT for authentication, which involves verifying the signature of incoming tokens and extracting the user ID and other claims from the token payload. To generate JWT tokens, we need to define a custom implementation of the Spring Security User Details Service interface, which retrieves user details from a data source such as a database or LDAP server. This implementation should generate a JWT token using a secret key and the claims for the user, and return the token to the client. Finally, we need to add unit and integration tests to ensure that the API is functioning correctly and securely. This involves creating test classes that use the Spring Test framework to simulate incoming requests and verify that the responses are correct.

Throughout the development process, it is important to follow best practices for secure coding, such as using parameterized SQL queries to prevent SQL injection attacks, validating user input to prevent cross-site scripting (XSS) attacks, and logging errors and exceptions to help with debugging and auditing.

Overall, developing a backend application using Spring Boot, JWT, and Maven involves a number of steps, each of which is essential for building a secure and reliable API. By following best practices and using the right tools and frameworks, developers can create high-quality APIs that are easy to use and maintain.

For deployment in cloud AWS is used Amazon Web Services (AWS) is a cloud computing platform provided by Amazon.com. It offers a wide range of services including computing power, database storage, content delivery, and other functionality to help businesses and organizations scale and grow.

One of the key benefits of AWS is its scalability. With AWS, businesses can easily scale up or down their computing resources depending on their needs, without having to worry about buying and maintaining expensive hardware. This allows businesses to save money and focus on other important aspects of their operations.

Another benefit of AWS is its reliability. AWS has a highly redundant infrastructure that is designed to ensure high availability and uptime for its services. This includes data centers located in multiple regions around the world, as well as automated failover and backup systems that help to ensure continuous operation of services.

AWS also offers a wide range of services that can be used fora variety of purposes. For example, businesses can use AWS to host their websites, run big data analytics, store and share files, build and deploy applications, and much more.

In addition to its core services, AWS also offers a range of tools and resources to help businesses get the most out of their cloud computing infrastructure. This includes tools for managing and monitoring resources, as well as training and certification programs to help businesses develop the skills they need to use AWS effectively.

Another important aspect of AWS is its security. AWS has a number of security features built into its platform, including encryption of data at rest and in transit, as well as access control and monitoring tools to help businesses protect their data and systems from unauthorized access.

Overall, AWS is a powerful and flexible cloud computing platform that offers a wide range of services and tools to help businesses scale and grow. With its scalability, reliability, and security features, AWS is an ideal choice for businesses looking to take advantage of cloud computing technology to improve their operations and grow their businesses.

7. Conclusion

This paper proposed a secure medical resource management system having secure transaction architecture. This provides a reliable and robust user- friendly secure system that allows the medical institutions to work collectively and increases the chances of patients' survival. This system will to store the data of the transactions and other details that will be beneficial in solving other problems. By using this system, we can overcome the problem of inefficient use of medical resources. This system is very efficient and allows medical institutions to manage and share their resources.

In conclusion, a medical resource management system isa crucial tool for healthcare organizations to manage their resources efficiently and effectively. By implementing such a system, healthcare facilities can optimize their resource utilization, reduce waste, improve patient care, and increase their overall operational efficiency.

A well-designed medical resource management system should have features such as real-time inventory tracking, automated order processing, demand forecasting, and analytics and reporting capabilities. It should also be user- friendly and integrate seamlessly with existing electronic health record systems.

Additionally, a medical resource management system can be customized to fit the unique needs of different health care facilities, depending on their size, specialty, and patient volume. Therefore, it is important for healthcare organizations to evaluate their specific resource management needs and choose a system that can meet those needs effectively.

Overall, a medical resource management system can help health care organizations better manage their resources, streamline their operations, and ultimately provide better care to their patients.

References

1. Verma, S.K., Manhas, P.S., Gupta, A.K., Kishore, V.V.N., and Devotta, S. (2020). A comprehensive review on the management of healthcare waste. Journal of Environmental Chemical Engineering, 8(5), 104162. doi: 10.1016/j.jece.2020.104162

2. Umar, R., Naeem, M.A., Ali, M., and Siddiqui, M.T.H. (2021). A review of healthcare waste management practices in low- and middle-income countries. Waste Management & Research, 39(3), 260-271. doi:10.1177/0734242X21988244

3. Azahar, B., Aziz, H.A., Nizam, M., and Adnan, M.A. (2019). An overview of medical waste management in developing countries: challenges and opportunities. Journal of Material Cycles and Waste Management, 21(5), 968-981. doi: 10.1007/s10163-019-00915-3

4. Alqahtani, A., Ibrahim, A.M., and Mahmoud, N. (2021).Medical waste management: a review. Journal of Environmental Management, 294, 112981. doi: 10.1016/j.jenvman.2021.112981

5. Alshehri, A., Javed, A., and Al-Ansari, T. (2020). A review of medical waste management in the Middle East and North Africa region. Waste Management & Research, 38(9), 978-990. doi:10.1177/0734242X20953970

6. Kumar, R., and Singh, R. (2021). Medical waste management during COVID-19 pandemic: a review. Environmental Science and Pollution Research, 28(3), 2941-2959. doi: 10.1007/s11356-020-11589-1